OUR SOLAR SYSTEM

EARTH
AND MOON

Robin Kerrod

Thameside Press

Distributed in the United States by
Smart Apple Media
1980 Lookout Drive
North Mankato, MN 56003

Text copyright © Robin Kerrod 2000

Editor: Veronica Ross
Designer: Jamie Asher
Illustrator: David Atkinson
Consultant: Douglas Millard
Picture researcher: Diana Morris

Printed in Singapore

Library of Congress Cataloging-in-Publication Data

Kerrod, Robin.
    Earth and moon / by Robin Kerrod.
        p. cm. -- (Our solar system)
    Includes index.
    Summary: Describes the formation, structure, and atmosphere
  of the earth and its moon, with up-to-date color photographs.
    ISBN 1-929298-64-1
    1. Earth--Juvenile literature. 2. Moon--Juvenile literature.
    [1. Earth. 2. Moon.] I. Title.

QB631.4.K465 2000
525--dc21   00-024749

9 8 7 6 5 4 3 2 1

PHOTO CREDITS
Paul Chesley/Tony Stone Images: 18tr.
European Press Agency/PA News: 13b.
Robin Kerrod/Spacecharts: cover inserts, 1,3,4 NASA, 5t, 5b ESA, 11t
NASA, 12-13, 14,15,16, 17t, 18c, 19t, 20, 21b, 22t, 22b, 24-5 background,
25t, 26, 29t, 29b NASA, 30-1, 32, 33, 34 NASA, 35 Lick Observatory, 35b
NASA, 36l, 36br NASA, 37 NASA, 38 NASA, 39 NASA, 40 NASA, 41
NASA, 42-3 background.
Brad Lewis/Tony Stone Images: 9b.
Joe McBride/Tony Stone Images: 7t.
Alan Puzey/Tony Stone Images: 21t.
Eddie Soloway/Tony Stone Images: 25b.
Tony Stone Images: 10b.
Stuart Westmorland/Tony Stone Images: 17c.
Peter Willi/The Louvre/Bridgeman Art Library: 28t.

Some of the more unfamiliar words used in this book
are explained in the glossary on pages 46 and 47.

CONTENTS

The Earth is our home in space. Alien space-travelers visiting our solar system would see that it is quite different from the other planets. From a distance, they would see the Earth as a mainly blue sphere, flecked with white and brown, floating in the inky blackness of space. A beautiful place.

Moving closer, our alien visitors would find that the blueness of our planet is the vast oceans. The white patches are clouds, and the brown ones are land.

Long before they reached the clouds, they would see rivers and lakes, mountains and valleys, and networks of city streets. Their cameras would spy the tiny, two-legged creatures that inhabit the cities. They would find that these creatures—humans—are just one of millions of different kinds of living things on the planet. How different, they would think, is this colorful, living Earth from the drab Moon that travels with it through space.

PLANET EARTH

The Earth is one of the smallest planets in the solar system.

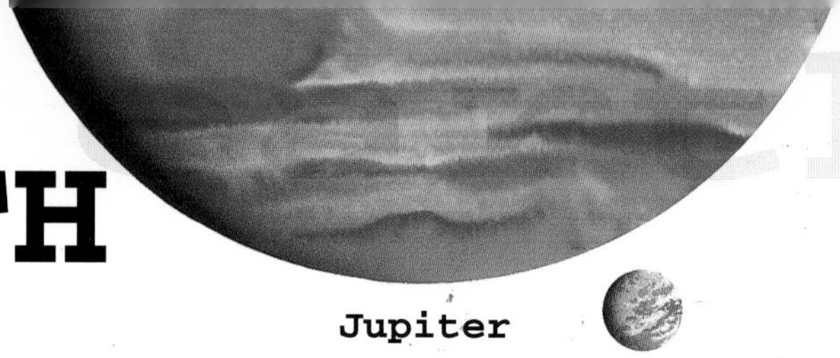

Until the 1500s, people believed that the Earth was at the center of the universe. All the other heavenly bodies—Sun, Moon, planets, and stars—circled the Earth. "Look at the Sun," they said. "Every day it circles the Earth."

So it appears, but things are actually the other way round. The Earth and planets circle the Sun, as Copernicus pointed out in 1543. The Earth is one of the smallest planets, and certainly not the center of, or the most important body in, the universe!

Jupiter

Earth

△ Earth is tiny compared with Jupiter

△ Astronomers such as Ptolomy thought the Earth was at the center of the Universe.

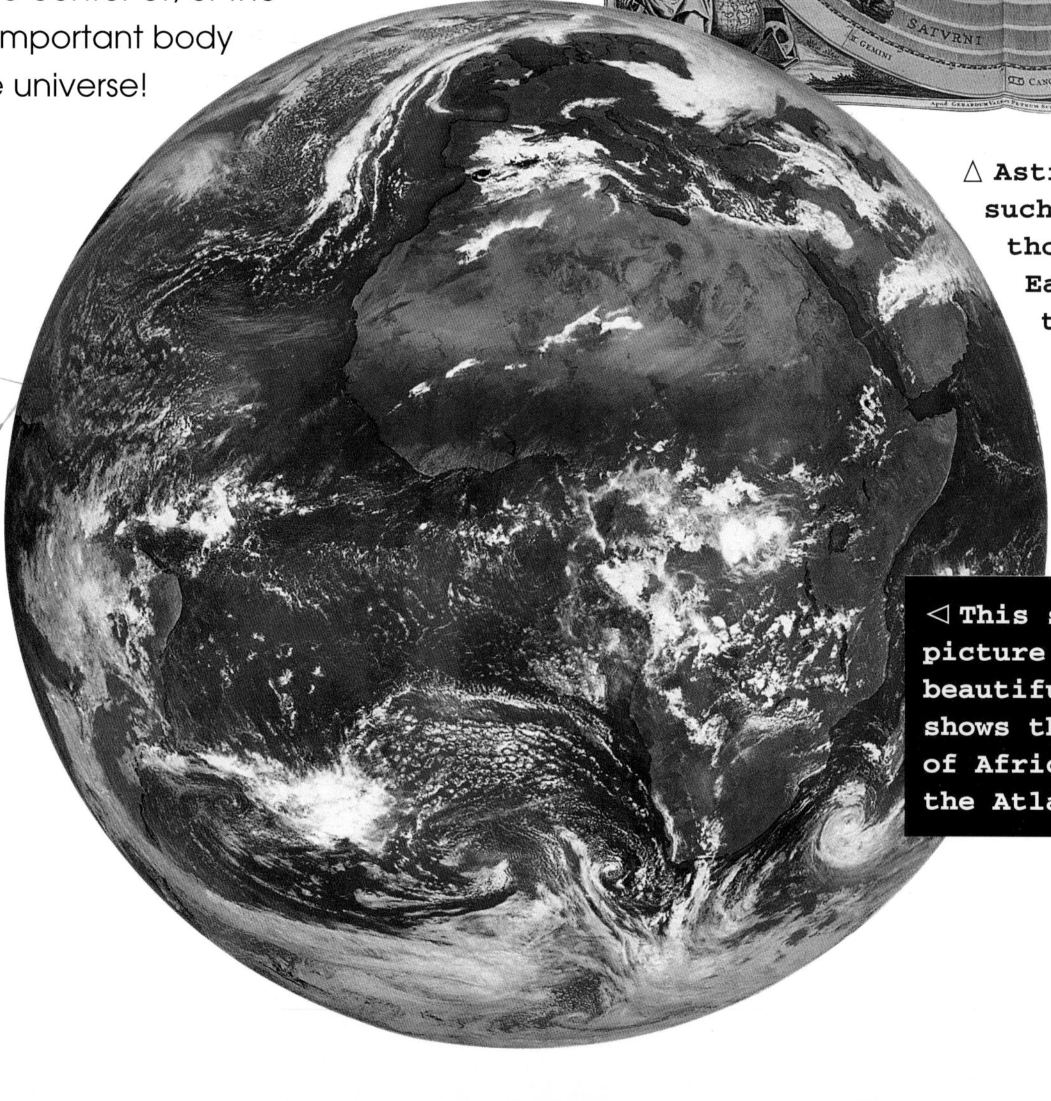

◁ This satellite picture of our beautiful Earth shows the continent of Africa and the Atlantic Ocean.

▷ Pulled by gravity, sky
divers fall to the ground
at frightening speeds.

Traveling through space

To us, it seems as if the Earth stands still, but
it is really rushing headlong through space.
Like the other planets, the Earth spins on its
axis and travels around the Sun in
a nearly circular path, or orbit.
It stays on average about
93 million miles from the
Sun. It spins round, like
a top, once every 24
hours—the time we
call a day. And it takes
365¼ days to travel
once around the Sun
—the time we call a year.
   The Earth's axis is not
upright, but is tilted in relation to
its path around the Sun. Parts of the Earth
tip more toward the Sun at some times
of the year than at others. These places
have regular changes in temperature
and weather at different times, or
"seasons," of the year.

◁ In the solar
system, the Earth
is the third planet
out from the Sun.

Earth's forces

Like all bodies, the Earth
has gravity. This is the
force of attraction, or pull,
it exerts on any object on
or near it in space. Gravity is the
force that makes things fall when you drop
them and keeps everything in place on the
Earth. Gravity extends into space, keeping
satellites and the Moon in orbit around the
Earth. The Earth is also magnetic. It behaves
as though it has a big bar magnet inside it,
with its two poles (ends) near the North
and South Poles. Earth's magnetism not
only affects things on the surface, such
as compasses, but also extends into space.
It forms a great magnetic "bubble" around
the Earth called the magnetosphere.

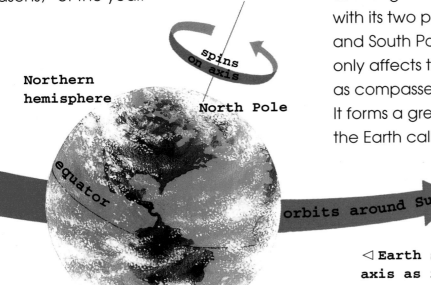

Northern
hemisphere

North Pole

spins
on axis

equator

orbits around Sun

South Pole

Southern
hemisphere

◁ Earth spins on its
axis as it circles in
space around the Sun.

FORMATION AND STRUCTURE

The Earth is made up of layers, rather like a gigantic onion.

The Earth was born at the same time as the other planets in the solar system, about 4,600 million years ago. It formed out of lumps of rock whirling in space around the early Sun. The lumps collided and stuck together to form the very large round mass we now call the Earth. The newborn Earth was very hot and molten (liquid) and took millions of years to cool down. As it cooled, gravity pulled heavy metals, such as iron and nickel, to the center, while lighter materials settled above it. This created a layered Earth. Geologists work out the structure of the Earth by tracing how earthquake waves travel through underground rocks.

▷ **How the Earth formed.**

1 Small lumps of matter stuck together to form large ones.

2 In time, a body the size of the Earth formed and became hot.

3 Later, gases coming out of the ground formed the early atmosphere.

4 After billions of years, Earth turned into the body we know today.

▷ **Inside the Earth there are different layers. The crust and mantle are made up of rock, but the core is metal.**

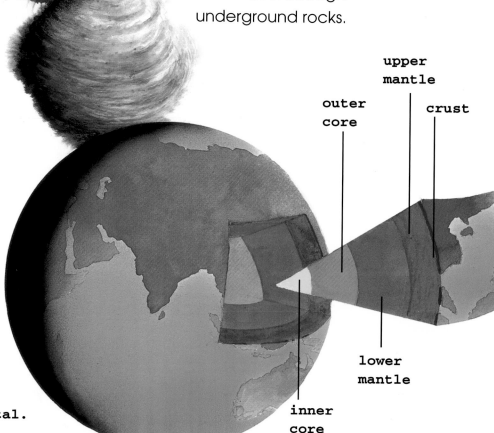

upper mantle

outer core

crust

lower mantle

inner core

paths of earthquake waves

mantle

inner core

outer core

shadow zone

△ **Earthquake waves bend when they pass between the layers of the Earth. Study of these waves is called seismology.**

The crust

Earthquake waves can suddenly change direction at certain depths underground. This tells geologists that the waves are entering layers of different materials. Using this information, they have discovered that the Earth is made up of four main layers—the crust, mantle, outer core, and inner core.

The crust makes up the Earth's hard outer layer, or skin, and is very thin compared with the other layers. Under the oceans the crust is only about 4⅓ kilometres thick. Under the land areas, or continents, it is much thicker—up to 25 miles.

The mantle and core

Under the crust lies a very deep layer called the mantle, made up of heavier rock than the crust. In the upper part of the mantle the rock is partly molten and moves slowly, carrying the crust with it (see page 10).

Underneath the mantle is the metal core, made up mainly of iron and nickel. The inner part of the core is solid, but the outer part seems to be liquid. This probably explains why the Earth is magnetic. Scientists know that moving metals (the Earth is spinning) set up electric currents, and that these produce magnetism.

◁ **Molten rock from the mantle sometimes forces its way to the surface.**

THE CRACKED CRUST

Gradual movements of the crust slowly but surely alter the face of the Earth.

Geologists used to think that the Earth's crust was like a solid shell, and that the continents had always been where they are today. But in the early 1900s, a German scientist named Alfred Wegener came up with a revolutionary new theory. The continents were drifting, carried along by movements in the Earth's crust. This theory became known as continental drift.

The theory supposed that long ago all the continents were joined together into a great supercontinent (Pangea) then gradually drifted apart. This explained something that people had realized for years—that the continents of South America and Africa look as if they would fit together like pieces of a jigsaw. The continental drift theory said that they were once joined. As they split apart, the sea flooded in, creating the Atlantic Ocean.

▷ **Australian kangaroos are marsupials. Marsupials are also found in South America, suggesting that these continents were once joined.**

▽ **The supercontinent Pangea began to break up about 200 million years ago. The continents gradually drifted to where they are today.**

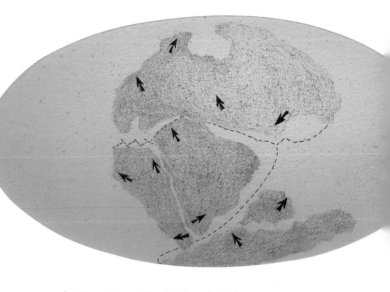

△ **The Earth 200 million years ago**

△ **The Earth 65 million years ago**

◁ **The Red Sea (center) is becoming wider because moving plates are forcing Africa (left) and Arabia (right) apart.**

The evidence

Most geologists did not accept the idea of continental drift until the 1960s. They have now found plenty of evidence to support it. South America, Africa, Antarctica, and Australia all have rocks and fossils of similar species (kinds) of reptiles and plants. This proves that they were once joined together in a supercontinent.

▽ **The Earth today. The lines show the boundaries between the plates that make up the crust.**

Theory of plates

A theory called plate tectonics explains how continental drift happens. It says that the Earth's crust is not solid, but is made up of a number of sections, or plates, which are moving. Some of these plates carry the continents.

The plates are made of solid rock. But the rocks beneath them, in the upper part of the mantle, are hot and semi-molten (partly liquid), like thick treacle. The rocks in the mantle flow slowly. They rise as they heat up, spread beneath the crust, then sink again as they cool. This is similar to the way warm air rises, cools, and sinks in a heated room. As the treacly rock flows beneath the crust, it carries the plates with it. In some places the plates are moving apart, and in others they are being pushed together. There are seven large plates and many smaller ones.

ON THE EDGE

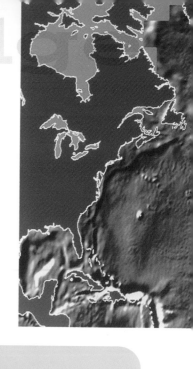

Volcanoes erupt and the Earth shakes where the plates of the crust meet.

The plates of the Earth's crust are moving apart under the oceans. For example, on the floor of the Atlantic, one plate (carrying South America) is moving towards the west, while the other (carrying Africa) is moving east. This is called sea-floor spreading.

As the two plates move apart, molten rock wells up from below to fill the gap and becomes part of the spreading plates. The boundary between spreading plates is called a constructive margin because new plate material is being created.

▽ **In mid-ocean, molten rock from below wells up and makes new plate material. At the ocean edges, plates descend and are destroyed.**

Black smokers
Along the mid-ocean ridges hot water escapes from the seabed. It is often colored black by dark minerals. The black streams look like spirals of smoke, and are called black smokers. Strange life forms exist around black smokers, including sulfur-eating bacteria, worms, and prawns.

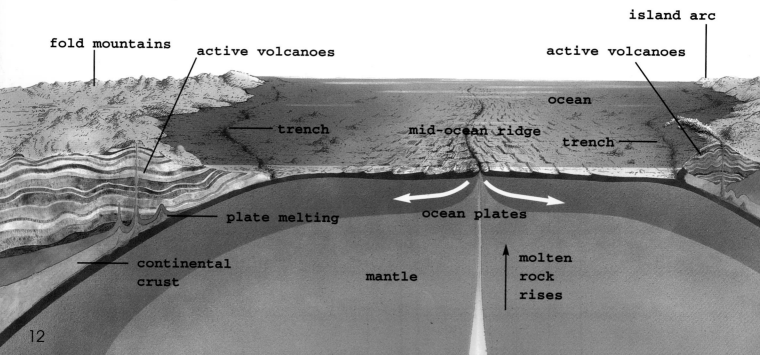

island arc

fold mountains   active volcanoes   active volcanoes

ocean

trench   mid-ocean ridge   trench

plate melting   ocean plates

continental crust   mantle   molten rock rises

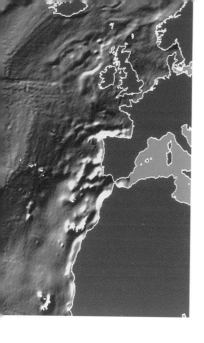

◁ This satellite picture centers on the Atlantic Ocean. It shows clearly the Mid-Atlantic Ridge.

The ocean ridges

Where molten rock wells up along a constructive margin, a mountain range called an ocean ridge grows up. The ridge that snakes across the floor of the Atlantic Ocean is called the Mid-Atlantic Ridge. For most of its length, it is hidden beneath the waves. But it rises to the surface in the far north, where it creates the island of Iceland. Here, molten material is constantly welling up from below, forming volcanoes.

▽ Devastation was caused by the 1999 earthquake in Turkey, in which tens of thousands of people died.

Plate destruction

There is also an ocean ridge in the Pacific, called the Pacific Rise. From the ridge, one plate spreads westwards and the other eastwards. The plate spreading eastwards collides with the plate carrying South America, which is moving westwards. The lighter South American, or continental, plate rides up over the ocean plate and forces it down. As it descends, the ocean plate melts and is destroyed. This kind of boundary is called a destructive margin.

Mountain building

The South American plate does not escape damage during its collision with the ocean plate. It crumples up to form the towering Andes Mountains. Deep below, the ocean plate melts and red-hot rock forces its way upwards. It spurts out of the surface, creating volcanoes.

As the plates grind against one another, they often lock together then suddenly become free. This sends shock waves through the rocks, causing earthquakes.

ROCKS AND MINERALS

Many rocks are born in fiery volcanoes or deep undergound.

The Earth's crust is made up of many different kinds of rocks. They vary widely in content, color, hardness and density. They are made up of different chemical compounds, or minerals. In some, the minerals are too small to be seen. In others, the minerals show up as colored specks or crystals.

  Quartz, or silica, is one of the most common minerals found in rocks. Quartz is the chemical silicon dioxide, a combination of the elements silicon and oxygen—the two most common elements in the Earth's crust. Many other minerals, called silicates, contain the same two elements and are found widely in rocks.

△ **The red-hot lava flows from volcanoes destroy everything in their path.**

▽ **Minerals are found in rocks in a variety of different shapes and colors. Galena (below) often forms shiny crystals with square faces. Malachite (below right) is a beautiful green color.**

Crystal clear

You can grow your own crystals from bath salts, washing soda, or alum, which you can buy from a drugstore. Add the substance to hot water and stir. Keep adding it until no more will dissolve. Let the solution cool. Dangle a piece of thread into it and leave for several days. Crystals will grow on the thread.

Volcanic rocks

Volcanoes form when hot molten rock, or magma, forces its way to the surface. The molten rock, now called lava, pours out and quickly cools. It sets hard, usually forming a dark-colored rock known as basalt. It is fine-grained, which means that its mineral crystals are too small to be seen.

Basalt is one kind of igneous, or fire-formed, rock. If lava cools very quickly, it becomes a kind of black glass called obsidian. If it contains lots of gas bubbles, it becomes a frothy rock called pumice, which can float on water. These rocks are called extrusive rocks because they form on the surface.

Intrusive rocks

Often, molten magma in the Earth's crust never makes it to the surface. It forces its way into cavities in the rocks and between rock layers and then stops. It takes ages to cool down, and during this time the minerals it contains grow into large crystals. This results in coarse-grained rock. The commonest rock of this kind is granite. Its crystals of white quartz, black mica, and pink felspar can be seen clearly. Rocks like granite that form inside the Earth are called intrusive rocks.

▽ This massive granite cliff, called El Capitan, is found in Yosemite National Park in California. It formed deep underground and appeared when the softer rocks around it eventually wore away.

RECYCLED ROCKS

Most surface rocks are made up of materials that settled out of ancient seas.

Rocks on the surface of the Earth are constantly being eroded, or worn away (see page 18). Rivers carry away the bits of rock and dump them in the sea. As layers of these deposits, or sediments, build up, the weight of the upper layers compresses the lower layers into hard sedimentary rock.

Sandstone is one of the commonest sedimentary rocks. It is made up of particles of sand naturally cemented together. Shale is a softer, flakier rock that was once mud. Some sedimentary rocks are made up of coarser fragments. Conglomerate, for example, contains rounded pebbles.

Sedimentary fuel

Coal is a sedimentary rock made from the plants that grew on Earth millions of years ago. Decaying plant material built up on the ground in ancient forests, and was squashed between mud and sand. The mud and sand turned to rock and the plant material turned into coal.

▽ Climbers struggle to climb Uluru (Ayers Rock) in the Northern Territory of Australia. Made up of sandstone, it is the world's biggest monolith, or single mass of rock.

▷ Here in the Rocky Mountains, layers show clearly in the sedimentary rock.

Chemical sediments

Seas contain many dissolved minerals, such as salt. When ancient salty seas dried up, thick layers of salt were left behind and eventually turned into rock salt. Similarly, some limestones formed when seas containing dissolved calcium carbonate dried up.

Fossil rocks

Other limestones are made of calcium carbonate from the remains, or fossils, of sea creatures such as corals. Corals build a limy skeleton around themselves when they grow, and it is left behind when they die. Chalk is made up of the remains of microscopic sea creatures, such as diatoms. Rocks formed in this way are called biogenic, meaning "produced by living things."

◁ This coral reef, full of life and color, could one day become solid rock.

Changed rocks

Rocks undergo change not only on the surface, but deep underground. When they come into contact with hot magma, the rocks become hot, and new minerals may form as they cool down. Underground rocks may be squeezed and reshaped by the pressure of rock movements, particularly in regions where the plates of the Earth's crust rub together (see page 13). Rocks changed by heat or pressure are called metamorphic, which means "changed form." Slate, formed from shale, and marble, formed from chalk and limestone, are both common metamorphic rocks. They are much harder than the substances from which they came.

17

UNDER THE WEATHER

From the moment they are born, rocks come under attack from all kinds of forces.

Eventually, all the great mountain ranges of the world—the lofty Alps and the mighty Himalayas—will become flat plains. Over millions of years, they will be worn away by natural forces. This process, called erosion, is forever reshaping the landscape, usually very slowly. Erosion by rain, wind, sunshine, and frost is called weathering.

△ ◁ **Erosion has created spectacular landscapes in Monument Valley (above) and the Grand Canyon (left), in Western states.**

Mechanical weathering

In mechanical weathering, the rocks are attacked physically. Rain wears away the soil and soft rocks such as chalk. The wind can pick up sharp particles that can grind away, or sandblast, the rock surface. Sunshine sets up stresses in the rocks.

It heats the top layer of rock, while the layers below remain cold. The top layer tries to expand and eventually peels off. Frost shatters rock. Water that has seeped into cracks in the rocks freezes into ice. Ice takes up more room than water. It forces the cracks apart until the rock shatters.

emerging stream ⟶

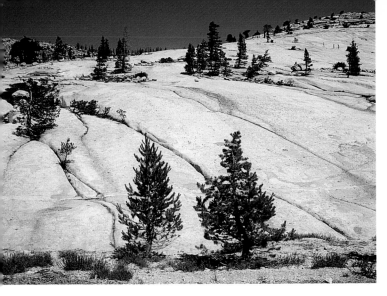

△ **In the Californian Sierras, trees grow out of the rocks. They too play a part in erosion. Their roots enter tiny cracks in the rocks and make them bigger and bigger, until the rocks split.**

The Grandest Canyon

Arizona's Grand Canyon is perhaps the finest example of erosion found on Earth. It is a huge valley more than 186 miles long, 4-18 miles wide, and over 1 mile deep in places. The canyon was formed by the Colorado River cutting through the landscape for a million years or more.

Chemical weathering

Rocks are also attacked chemically, mainly by substances dissolved in rain and river water. Rainwater naturally absorbs carbon dioxide from the air, which makes it slightly acid. Sulfuric acid from air pollution makes it more acid. The acid water runs over the rocks and dissolves some of the minerals in them.

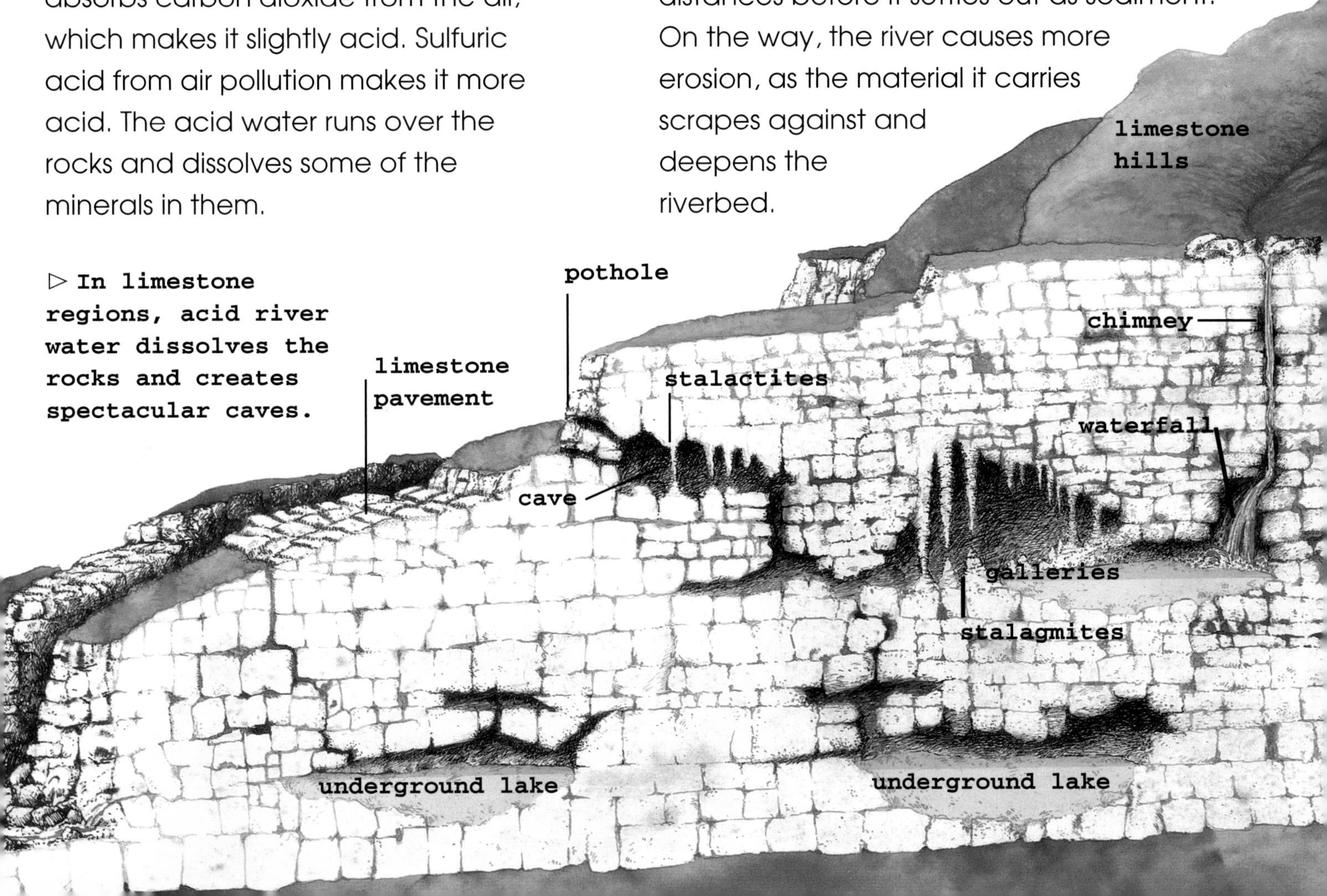

Transport

Weathering is only part of the erosion process. Rivers transport the weathered material. They can carry it for long distances before it settles out as sediment. On the way, the river causes more erosion, as the material it carries scrapes against and deepens the riverbed.

▷ **In limestone regions, acid river water dissolves the rocks and creates spectacular caves.**

limestone hills
pothole
chimney
limestone pavement
stalactites
waterfall
cave
galleries
stalagmites
underground lake
underground lake

PLANET WATER

On Earth there is water in the oceans, on the land, and in the air.

More than two-thirds of the Earth's surface is covered with water—the water of the oceans. All the land in the world would fit into the largest ocean, the Pacific.

On average, the oceans are about 2½ miles deep, but in some places they plummet to more than 7 miles. They are so vast and so deep that they contain 97 percent of the Earth's water. The oceans have a huge effect on the world's climate and weather. They absorb and store great amounts of the Sun's heat. They hold this heat better than land does, and so have a steadying effect on world temperatures.

As the Sun heats the oceans, water evaporates from the surface and escapes into the air as water vapor. Eventually, the vapor turns back to liquid and falls to the surface as rain or snow in a never-ending process called the water cycle (see page 24).

Great Lakes

The five Great Lakes form the largest body of freshwater in the world. Together they cover an area of 95,500 square miles, which is about half the size of France. From largest to smallest, they are Lakes Superior, Huron, Michigan, Erie, and Ontario.

△ From space, the Earth has a watery image, with vast oceans of liquid water, clouds of water vapor and sheets of water ice.

Salty and fresh

Seawater is very salty. It contains more than three parts in a hundred of ordinary salt (sodium chloride) and other salts. Rivers dissolve tiny amounts of salts from the rocks they run over and carry them to the sea, where they collect.

River water contains very little salt and so is called freshwater. There are also large amounts of freshwater in lakes and underground rocks, where it is called groundwater. The deepest lakes are in northern North America and Europe. They were dug out of the landscape by glaciers during the last ice age.

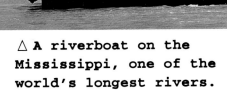

△ A riverboat on the Mississippi, one of the world's longest rivers.

Icy wastes

There are still many glaciers on Earth today, in high mountain ranges such as the North American Rockies and the Alps in Europe. They too are a store of freshwater. Even more freshwater is locked in the permanent ice fields that exist in Antarctica around the South Pole, and in the Arctic around the North Pole.

▷ If you could sink Mount Everest in the deepest part of the oceans, its summit would still be more than 1 mile below the surface.

THE BLANKETING ATMOSPHERE

A thin covering of air warms and protects the Earth.

The outermost layer of the Earth, called the atmosphere, is made up of gases. The atmosphere makes life on Earth possible. It lets through the heat of the Sun by day, and traps enough heat to keep our planet warm at night. The atmosphere shields us from harmful rays that come from the Sun and outer space. They include X-rays and the ultraviolet rays that cause sunburn.

The vital gases

The atmosphere also provides vital gases. Together, the gases in the atmosphere are called air. Air is made up of nitrogen (78 percent), oxygen (21 percent), and argon (less than 1 percent). Oxygen is vital for life, because almost all living things need it for respiration, or breathing—plants as well as animals. Carbon dioxide is another vital gas, which plants use to make food. If plants could not make food, there would be no food for animals because animals cannot make their own food.

△ Weather satellites spot changes in the atmosphere.

▽ An astronaut's view of the atmosphere just before sunrise.

Layers of atmosphere

The atmosphere is held to the Earth by gravity and has weight. The air is thickest, or densest, near the surface. There, a weight of 13.8 lb. of air presses down on every square inch. This is called atmospheric pressure. The air becomes thinner, or less dense, and the pressure falls as you go higher above the surface. There are layers in the atmosphere, each with different properties. Most of our weather takes place in the lowest, thickest layer, called the troposphere. Above the troposphere the air thins rapidly.

Global warming

The amount of carbon dioxide in the atmosphere is increasing as people burn more fuel in power stations and cars. This gas makes the atmosphere trap more heat. It is causing the world's climate to warm up. Scientists fear that this global warming will upset world weather systems.

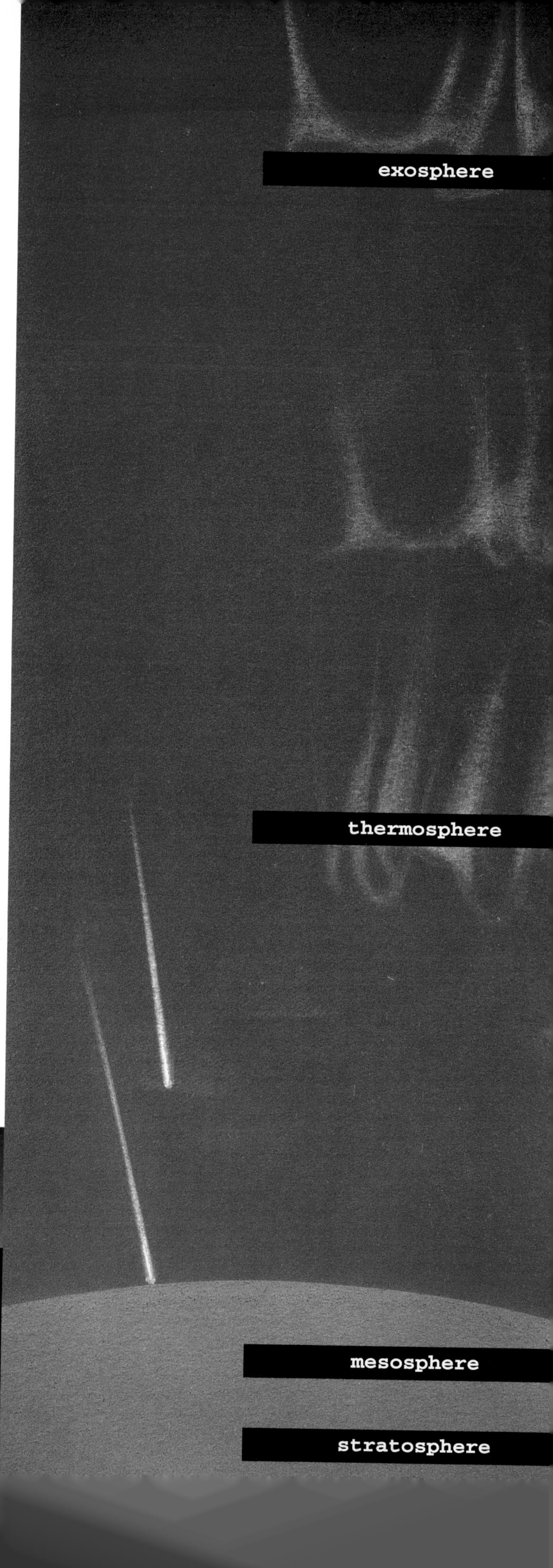

exosphere

thermosphere

▷ The layers of atmosphere around the Earth. We live in the troposphere, the lowest and densest layer. It is nearly 7 miles thick near the North and South Poles.

mesosphere

stratosphere

WATER CYCLE

Water circulates constantly between the surface and the atmosphere.

Much of our weather depends on the amount of moisture, or water, there is in the air. If there is a lot of moisture, it will probably rain—or snow if the temperature is low. Or, it may become foggy.

Water escapes from the Earth's surface into the air all the time. Much of it comes from the oceans. The Sun heats up the water and makes it evaporate, or turn into vapor (gas). The water vapor rises into the air and mixes with the other gases. Trees and other plants also give off water vapor. They draw up water from the ground and use some to make their food. The rest escapes into the air. This process is called transpiration.

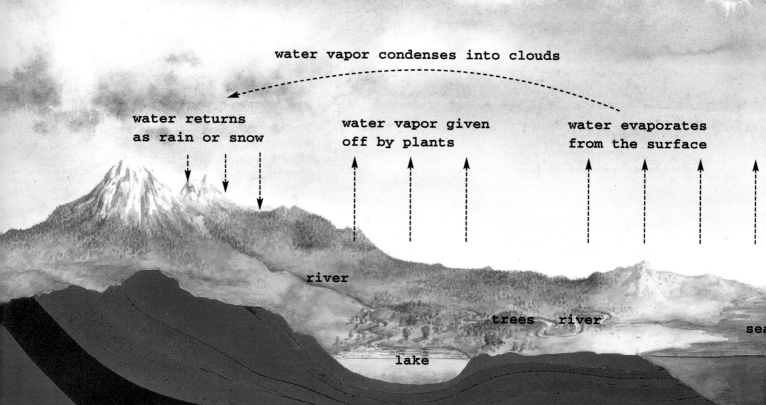

▽ The water cycle constantly recycles the Earth's water supply.

water vapor condenses into clouds

water returns as rain or snow

water vapor given off by plants

water evaporates from the surface

river

trees   river

lake

sea

△ **Snow drifts can be deep in the Appalachian Mountains.**

Cooling down

The higher up you go, the cooler the atmosphere. So as water vapor rises into the air, it cools down and turns back into tiny droplets of liquid water. Great masses of droplets gather together to form clouds.

Inside the clouds, the tiny water droplets collide with one another to form bigger and bigger drops. Eventually, the drops may become so heavy that they fall from the cloud as rain. If the air temperature is low enough, the drops freeze and fall as bundles of tiny ice crystals called snow.

On the ground, the rainwater or melted snow runs into rivers and lakes or into the oceans. As evaporation and transpiration continue, the water returns again to the air in a never-ending process known as the water cycle.

Precipitation

Rain and snow are two forms of precipitation, or ways in which water falls from the air to the ground. Water comes from the air in other ways, too. On cool nights, drops of water form on surfaces as dew. If temperatures are below freezing, the water vapor changes directly to ice crystals, creating a white frost.

Thunder and lightning

A thunderstorm is a natural fireworks display. Lightning occurs when millions of volts of electricity build up in thunderclouds. Suddenly the electricity discharges (leaks away) in an electric spark called lightning. The air around the lightning flash heats and expands so quickly that it creates an explosion called thunder.

△ **Lightning zigzags from clouds during a thunderstorm.**

LIVING PLANET

The Earth is the only place in the whole universe where we know life exists now.

◁ The Earth is home to an enormous number of animal species. Here, bottle-nosed dolphins perform acrobatics in the ocean.

The Earth is in just the right position in the solar system for life to thrive. It is not too hot, and not too cold. Water exists as a liquid, which is essential for all forms of life.

No one knows precisely how life began on Earth. Many scientists think that it began chemically. Substances in the Earth's early atmosphere, such as methane and ammonia, reacted together to form simple organic compounds. Organic compounds, containing chains of carbon atoms, hold the key to life as we know it.

The simple organic compounds fell into the warm oceans and began reacting together to form more complicated compounds. Eventually, compounds formed that could reproduce themselves. This was the start of life on Earth.

△ Bougainvillea dazzles the eye with stunning flowers. It is one of about 250,000 species of flowering plants. There are also thousands of species of non-flowering plants, such as ferns and mosses.

▽ **Plants make food by photosynthesis: leaves take in carbon dioxide, combine it with water to make food, and give off oxygen.**

△ Grass takes in carbon dioxide to make its food. Sheep eat grass, and the carbon in it combines with the oxygen they breathe in to produce carbon dioxide, which they breathe out.

△ **When wood burns, the carbon it contains combines with oxygen to give off carbon dioxide.**

Life develops

The earliest forms of life appeared in the oceans more than 3,000 million years ago. They were simple organisms, probably like the bacteria on the Earth today. Soon, organisms developed that had cells, as most living things do today. They included simple plants such as algae and worm-like creatures with soft bodies.

These early life forms left only a few traces in the rocks when they died. Much later, creatures with shells and other hard body parts left behind plenty of remains, or fossils, in rocks formed about 590 million years ago. This was at the beginning of a period known as the Cambrian Period.

As millions of years went by, new and more complicated life forms developed, and left fossils behind. By studying fossils and dating the rock layers in which they are found, scientists can trace the evolution (development) of life on the Earth from the earliest times to today.

Spores from space
Some astronomers think that life did not begin on this Earth, but came from space. They think that primitive organisms—life spores—were carried to Earth by meteorites or comets. They also think that these spores—organisms like bacteria and viruses—are still raining down on this Earth, causing diseases.

THE EARTH'S MOON

The Moon is Earth's closest companion in space.

When the Moon shines in the night sky, its silvery beams give us enough light to see. Long ago, people worshipped the Moon because it brought them light at night, just as they worshipped the Sun for bringing light during the day. The Romans called the Moon Luna, from which we get our word "lunar," meaning to do with the Moon.

The Moon is still a special place to us. It is the Earth's closest neighbor in space, never more than about 248,000 miles away. And it is the only other world in the Universe that humans have so far visited. Apollo astronauts began exploring the Moon in 1969, making the most exciting adventure in modern history.

△ **The Roman goddess Diana, whom the Greeks called Artemis. She was also goddess of the hunt, using the crescent Moon as her bow and moonbeams as her arrows.**

Earth

Moon

Ganymede

▷ **The Moon, compared in size with the Earth and Jupiter's moon Ganymede, which is the biggest moon in the solar system.**

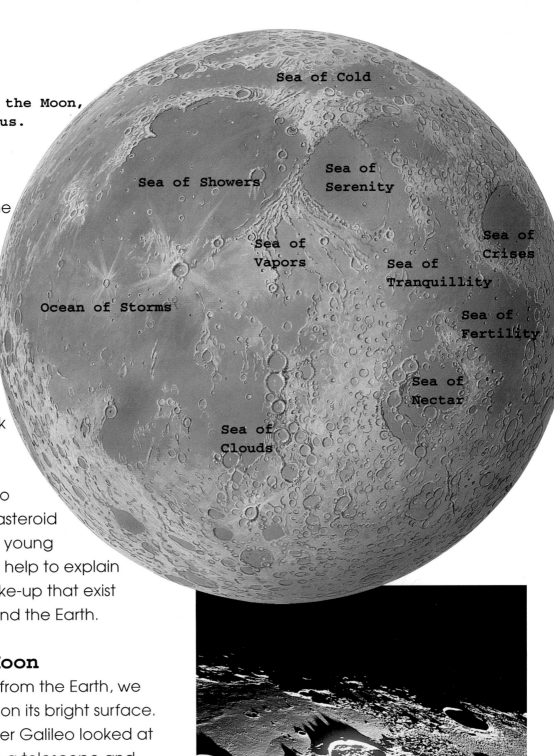

Sea of Cold

Sea of Showers

Sea of Serenity

Sea of Vapors

Sea of Crises

Sea of Tranquillity

Ocean of Storms

Sea of Fertility

Sea of Nectar

Sea of Clouds

Origins

No one is quite sure where the Moon came from. It might have formed at the same time as the Earth and the other planets, from the rocky lumps that were whizzing about in space. But most astronomers now think that the Moon was formed later, from material scattered into space when a huge asteroid collided with the very young Earth. This idea would help to explain the differences in make-up that exist between the Moon and the Earth.

Seas on the Moon

Looking at the Moon from the Earth, we can see dark regions on its bright surface. In 1609, the astronomer Galileo looked at these regions through a telescope and thought that they looked like the seas we have on Earth. So he called them *maria*, which is Latin for seas. He likened the brighter regions to land masses and called them *terrae*, Latin for land.

We now know that there are no watery seas on the Moon, but we still call the dark regions seas. In reality, they are dusty plains, while the brighter regions are highlands.

△ **Part of the Ocean of Storms. The large crater is Kepler, which is about 22 miles across.**

CIRCLING THE EARTH

The waxing and waning of the Moon mark one of the great rhythms of nature.

The Moon is the Earth's only natural satellite, or moon. It travels around the Earth in space in a nearly circular path, or orbit. It circles the Earth once every 27⅓ days. During this time, it also spins round once on its own axis like a top. This causes the same side of the Moon to face the Earth all the time, and we call it the near side. From the Earth, we can never see the opposite, or far side, of the Moon.

Ever changing

If we look at the Moon night after night, it seems to change shape. Some nights it appears only as a thin crescent, and other nights it appears as a full circle. These apparent changes in shape are called phases. The phases of the Moon come about because it does not give off light of its own. We see it shine only because it reflects light from the Sun.

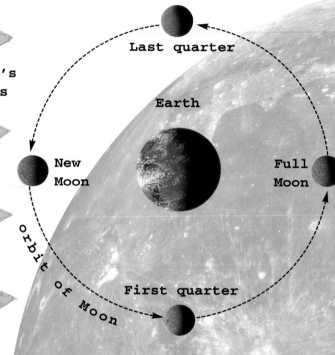

▽ We see different amounts of the Moon's surface lit up by the Sun at different times.

Last quarter

Sun's rays

Earth

New Moon

Full Moon

orbit of Moon

First quarter

▽ Five stages in the waxing and waning of the Moon, as the Moon goes through its phases.

Crescent Moon, 4 days old

First quarter, 7 days old

Full Moon, 14 days old

◁ **The far side of the Moon consists mostly of rugged highlands, with few seas.**

The phases

As the Moon circles the Earth, we see different amounts lit up by the Sun at different times. The amount we see depends on the positions of the Moon and Earth in relation to the Sun. When the Moon moves into line between the Sun and the Earth, the side facing us is dark and we cannot see the Moon at all. We call this phase the new Moon.

In a day or so, the Sun lights up one edge of the Moon, and we see it as a crescent. Each night, more of the Moon's surface is lit up, until after about a week the Moon appears as a half-circle (first-quarter phase). After another week, the Moon is a full circle (full-Moon phase). While it is growing, we say the Moon is waxing. After the full Moon, the Moon seems to shrink, or wane. After a week, it has shrunk to a half-circle (last quarter). After another week, it is a crescent before it disappears at the next new Moon.

Crescent Moon, 27 days old

Last quarter, 22 days old

SMALL WORLD

The Moon is a silent world of extreme temperatures and never-changing landscapes.

The Moon is a rocky body like the Earth, but is much smaller. Its diameter is only 2,155 miles, about a quarter that of the Earth—but this is relatively large for a moon. There are only four larger moons in the whole of the solar system.

In the sky, the Moon appears as large as the Sun and this brings about some of the most dramatic astronomical happenings —eclipses. Once or twice a year, the Sun, the Moon and the Earth move exactly into line in space.

▷ **During a lunar eclipse, the Moon turns pink or orange.**

Into the shadows

When the Moon comes between the Sun and the Earth, it casts a shadow on the Earth, causing an eclipse of the Sun. The Moon is quite small, so its shadow covers only a small area at a time. And at any place, the eclipse never lasts for more than a few minutes. When the Earth comes between the Sun and the Moon, the Moon enters the Earth's shadow, causing an eclipse of the Moon. Because the Earth casts a large shadow in space, the Moon can stay in eclipse for up to 2½ hours.

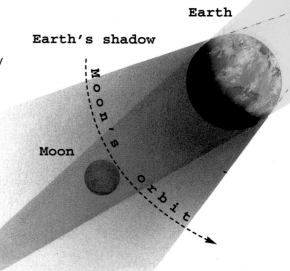

Earth

Earth's shadow

Moon's orbit

Moon

Weak gravity

The Moon is not only smaller than the Earth, it is also less dense. This means that it has a much weaker pull, or gravity, than Earth—about one-sixth as strong. But, it still affects the Earth by tugging at the water in the oceans, causing the tides.

Because of its weak gravity, the Moon cannot hold on to any gases to form an atmosphere. This has made it a quite a different world from the Earth. For one thing, it is completely silent, because sound waves cannot travel where there is no air.

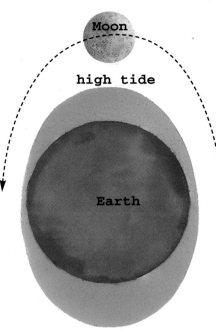

▷ **The Moon's gravity tugs at the Earth's seas, causing the rise and fall of the tides.**

Moon

high tide

Earth

low tide

How a lunar eclipse occurs: the Moon moves into the Earth's shadow in space.

Sun

Different worlds

On Earth, the sky is blue; on the Moon, it is black. On the Moon, the temperature varies enormously. During the lunar day, when the Sun shines, the temperature soars to more than the boiling point of water (212°C). During the lunar night, the temperature drops to -286°C below freezing. On Earth, flowing water and the weather are constantly changing the landscape. The Moon has no flowing water or weather, and the landscape has hardly changed for hundreds of millions of years.

▷ **Fishing boats are beached when the tide goes out.**

HIGHLANDS AND LOWLANDS

The Moon is a world of great plains and towering mountain ranges.

The dark areas on the Moon are vast plains, which we call seas. The seas cover more than half of the side of the Moon that faces us. Biggest by far is the Ocean of Storms, which is about the same size as the Mediterranean Sea on the Earth.

In the north, the Ocean of Storms merges into the circular Sea of Showers. On the western side of this sea is a large inlet known as the Bay of Rainbows. Early astronomers were fond of giving such delightful names to the Moon's plains regions. The two largest seas on the eastern half of the Moon are named Serenity and Tranquillity. It was near the southern edge of the Sea of Tranquillity that humans first planted their footsteps in the lunar soil (see page 40).

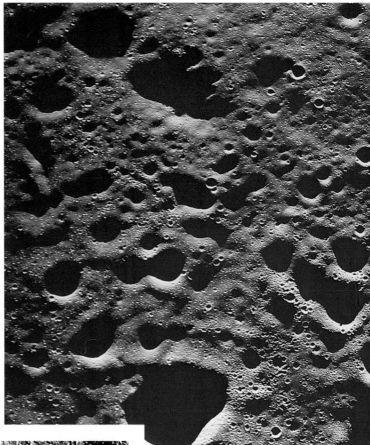

△ **The far side of the Moon is rugged and covered with ancient craters.**

△ **Alpine Valley, in the lunar Alps, runs into the Sea of Showers.**

In the highlands

The paler and brighter highland regions seem to be part of the Moon's original surface. They are older than the seas, and much more heavily cratered. The highest regions on the Moon are ranges of mountains ringing the large seas. They were probably thrown up by the impact of the asteroid-type bodies that created the huge craters that became the seas.

The Sea of Showers is ringed by mountain ranges that stretch for hundreds of miles and have peaks that rise to 19,680 feet. In the lunar Alps, Alpine Valley has a channel running through it that looks like a riverbed.

△ **The Apennine Mountain range curves around the Sea of Serenity.**

Rills and ridges

Most of the seas are circular in shape, and there is a good reason for this. Early in its history, the Moon was bombarded by enormous lumps of rock that gouged out huge craters. Later, volcanoes erupted in these craters and filled them with lava to form the seas we find today.

The surface of the seas still show evidence of volcanic activity. Many small craters were once volcanoes. Elsewhere, the surface has blistered into domes where lava has tried to push up from below. Long, snaking trenches, or rills, show where underground tunnels carrying lava emptied and collapsed.

▽ **The smooth surface of the Sea of Tranquillity, snapped by astronauts as they came in to land.**

COUNTLESS CRATERS

Since the day it formed, the Moon has been bombarded by meteorites, or rocky lumps from space. When these lumps smash into the surface, they dig out hollows, or holes, called impact craters. Most of the craters on the Moon are impact craters. Others are the mouths of ancient volcanoes.

Many of the craters we see today were formed billions of years ago. They have changed little over the years because there is no weathering and erosion to wear them away, as there is on Earth (see page 18).

The seas and highlands are peppered with craters, large and small.

The ice craters
In 1998, a space probe named *Lunar Prospector* found water ice in craters near the Moon's north and south poles. Scientists estimate that the craters contain thousands of millions of tons of ice, which could provide drinking water for Moon bases of the future.

▽ **The crater Eratosthenes, on the Ocean of Storms. On the horizon right is the larger Copernicus.**

△ **At the time of the full Moon, bright, shining rays surround the crater Tycho, in the far south.**

◁ This typical large crater has terraced walls and central mountain peaks.

Shapes and sizes

The craters of the Moon vary greatly in shape and size. Some are only little dimples in the surface, while others are deep and measure hundreds of miles across.

The crater Copernicus, in the Ocean of Storms, is a typical large crater. It measures about 56 miles across, is circular in shape, and has walls that rise on the outside hundreds of feet above the surrounding landscape. On the inside, the walls descend in terraces a mile or two below the ground level to a flat floor. In the center is a range of mountains.

A few large craters have fairly low walls and a smooth flat floor, and are generally called walled plains. Plato, near the northern edge of the Sea of Showers, is particularly noticeable because of its very dark floor.

▷ Craters Aristarchus (left) and Herodotus on the Ocean of Storms.

Old and new

Many of the older craters have changed over the years. In some, the walls have collapsed, while others have been hit by meteorites. On the surface of the seas, "ghost" craters can often be found. They are ancient craters that were flooded with lava when the seas formed, leaving just their tops showing.

From Earth, the easiest craters to spot are some of the newest. They include Tycho and Copernicus, which stand out clearly at full Moon because bright rays surround them. Crater rays are made by lines of glassy substances thrown out when the crater formed.

INSIDE THE MOON

Under a dusty crust strewn with rocks, the Moon is made up of layers.

When meteorites hit the surface of the Moon, they not only dig out craters, but also break up the surface rocks. Over billions of years, this process has ground down the rock and created a kind of soil called regolith.

Regolith is found everywhere on the Moon. It consists of fine particles of dust, rock fragments, and also tiny beads of glass. The beads form when a meteorite hits the surface hard and melts it. Tiny bits splatter up and quickly cool, forming a glassy substance. When the Apollo astronauts went to the Moon, the glass beads made the surface slippery for walking.

△ Edward Aldrin on the Sea of Tranquillity. His footsteps show clearly in the soft Moon soil.

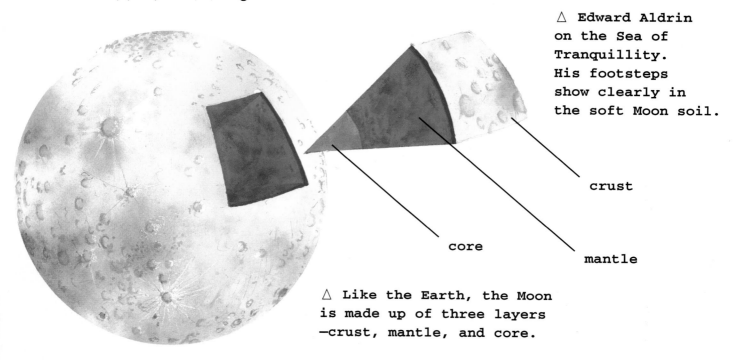

crust

core

mantle

△ Like the Earth, the Moon is made up of three layers —crust, mantle, and core.

△ **A scientist checks readings from the seismometers left on the Moon by the Apollo astronauts.**

Waves in the rocks

One of the Apollo astronauts' main tasks was to set up instruments called seismometers to detect moonquakes. Moonquakes are shakings in the lunar rocks similar to earthquakes on the Earth. The seismometers recorded the way moonquake waves traveled through the rocks. From this data, scientists worked out what the Moon is like inside.

Underneath the lunar soil is hard rock, which forms the Moon's outer layer, or crust. It is about 37 miles thick. In the upper part, the rocks seem to be riddled with cracks. Beneath the crust is a deep rock layer, called the mantle. This extends to a depth of about 620 miles and surrounds the Moon's center, or core. The outer part of the core may be partly molten, as it is on Earth.

Volcanic rocks

All the rocks the astronauts brought back from the Moon are igneous, or volcanic rocks. There are no sedimentary rocks because there have never been seas or rivers on the Moon to create them.

The dark rock that is typical of the Moon's seas is very similar to the basalt found on Earth. The typical rock of the lunar highlands is lighter in color because it contains more aluminum and calcium. Both kinds of rocks have a slightly different composition from similar rocks on Earth.

Rocks called breccias are found both on the seas and in the highlands. They are made up of rock fragments which have been "glued" together by glassy material thrown out from craters.

▽ **Two common kinds of rocks found on the Moon. Breccia (top) is made up of bits of old rock cemented together. Basalt (bottom) came from ancient volcanoes.**

THE ASTRONAUTS' MOON

Between 1969 and 1972, 12 astronauts left footprints on the Moon.

The Apollo project to land U.S. astronauts on the Moon began in earnest in 1961. President Kennedy set the goal "before the decade is out, of landing a man on the Moon and returning him safely to Earth." The goal was achieved, with five months to spare. The first Moon landing, by the *Apollo 11* spacecraft, took place on July 20, 1969.

▷ **It is July 16, 1969. The Saturn rocket fires to lift the *Apollo 11* crew from the Earth.**

Famous first words

When Neil Armstrong first stepped on to the Moon from *Apollo 11*, he meant to say: "That's one small step for a man, one giant leap for mankind." In the event, he left out the "a," which gave the sentence less impact. But how nervous and excited he must have been, watched by 500 million people on Earth.

The hardware

The Apollo spacecraft consisted of three parts, or modules. The crew of three traveled in the command module, which was linked for most of the time to the service module. The third part was the lunar module, which took astronauts down to the Moon. Altogether, the spacecraft weighed about 50 tons.

To launch such a craft required a powerful and gigantic new rocket, the Saturn V. On the launch pad, the rocket stood 364 feet high. It blasted off with the power of a fleet of jumbo jets. The Saturn V boosted the Apollo spacecraft, first into orbit around the Earth and then on to the Moon.

▷ *Apollo 17* astronaut Harrison Schmitt checks out a huge boulder during the last Moon-landing mission in December 1972.

The landings

When the Apollo spacecraft reached the Moon, it fired braking rockets and went into orbit. Two of the crew then descended to the Moon in the lunar module, leaving the command and service modules (CSM) still in orbit.

On the Moon, the astronauts picked up rock and soil samples and drilled deep into the ground. They carried out experiments and set up scientific equipment to record information and radio it back to scientists on Earth. They also took photographs of the extraordinary lunar landscape.

On the first three Moon landing missions (*Apollo 11, 12,* and *14*), the astronauts explored sea sites. On the last three missions (*Apollo 15, 16,* and *17*), they explored highland and valley sites. On these three missions, they were able to explore larger areas because they had a Moon buggy to transport them.

△ An astronaut rides the Moon buggy. Powered by electric motors, it had a top speed of about 9 mph.

TIME LINE

MYA stands for million years ago

4,600 MYA
The Earth is born, along with the other bodies in the solar system, from a huge cloud of gas and dust in space.

3,500 MYA
The first primitive life forms appear in the oceans.

590 MYA
By now, an explosion of life has taken place, recorded in the first abundant fossils in the rocks at the beginning of the Cambrian Period of Earth's history.

400 MYA
The first plants appear on land, while the first fish appear in the seas.

300 MYA
Huge fern forests now grow, which will later turn into the fossil fuel we know as coal. Amphibians and insects abound.

200 MYA
All the continents of the Earth come together to create the supercontinent Pangea, which later splits up. There is now only one huge ocean, called Panthallassa.

150 MYA
By now, dinosaurs are the dominant life form on the Earth, reaching huge dimensions.

65 MYA
Some planetwide, cataclysmic event occurs, maybe bombardment by asteroids, which wipes out the dinosaurs and many other species. Afterward, mammals evolve as the dominant life forms.

4 MYA
Our earliest ancestors grow up in what is now East Africa.

35 000 B.C.
The modern human race, Homo sapiens, begins to flourish.

3000 B.C.
The early civilizations believe the Earth is flat, and if you go too far, you will fall off.

300s B.C.
The idea is now firmly established among Greek philosophers like Aristotle that the Earth is spherical, and not flat.

200s B.C.
The Greek mathematician Eratosthenes devises a method of measuring the size of the Earth. His estimate of the diameter is only about 77 miles less than the true value (7,909 miles).

A.D. 79
The volcano Vesuvius erupts, burying the prosperous Roman town of Pompeii, near Naples, and killing thousands of its inhabitants.

150

The astronomer Ptolemy of Alexandria, now in Egypt, sets out in his book *Almagest* the accepted idea that the Earth is the center of the Universe, around which the other heavenly bodies revolve.

1543

Nicolaus Copernicus publishes a book in which he says that the Earth is not the center of the universe, but an ordinary planet that circles around the Sun.

1600

English scientist William Gilbert shows that the Earth is like a huge magnet and establishes the laws of magnetism.

1650

Irish archbishop James Ussher reports that, by studying the Bible, he has worked out that the Earth was created on October 23, 4004 B.C. This date for the creation is widely accepted until the 1800s.

1687

English scientist Isaac Newton publishes his law of gravity in his historic work, *Principia*.

1799

English scientist William Smith is first to use fossils as a means of dating rocks.

1815

Ash pumped into the atmosphere when the Indonesian volcano Tambora erupts cools the climate so much that the eastern U.S. has severe frosts in August.

1830

English scientist Charles Lyell publishes the first of three volumes of *Principles of Geology*, which puts the study on a sound scientific footing. He is one of the first to realize that the face of the Earth changes slowly, by natural processes which have been going on for thousands of millions of years.

1883

The explosion of the volcano Krakatoa in Indonesia unleashes a tidal wave that kills 36,000 people

1912

German scientist Alfred Wegener puts forward the theory of continental drift.

1931

U.S. scientist Charles Richter works out the Richter scale for measuring the strength of earthquakes.

1959

The Russian space probe *Luna 2* becomes the first man-made object to reach the Moon. Later, *Luna 3* sends back photographs of the far side, never seen before.

1961

In April, Russian pilot Yuri Gagarin becomes the first person to orbit the Earth, in the spacecraft *Vostok 1*.

1969

In July, *Apollo 11* makes the first landing on the Moon.

1972

In December, *Apollo 17* makes the last manned Moon landing so far.

1998

The space probe *Lunar Prospector* discovers water ice in the polar craters of the Moon.

1999

A violent earthquake in Turkey kills up to 40,000 people.

NOTHING BUT THE FACTS

Earth data

Diameter	at equator:	7,909 miles
	at poles:	7,882 miles
Circumference	around equator: 24,846 miles	
Surface area	total: 199 million square miles	
	land: 57 million square miles (29%)	
	ocean: 142 million square miles (71%)	

Volume: 108,300 cubic miles
Density (water's density =1): 5.5
Mass: 6,600 million million million tons

Distance from Sun	average: 92,752,000 miles
	furthest: 94,240,000 miles
	closest: 91,110,000 miles

Spins on axis in:	23 hours 56 minutes 4 seconds
Circles Sun in:	365.256 days
Speed in orbit:	66,340 mph

Moon data

Diameter at equator: 2,155 miles

Volume (Earth's volume =1): 1/49
Density (water's density =1): 3.3
Mass (Earth's mass =1): 1/81

Distance from Earth	average: 238,020 miles
	furthest: 252,340 miles
	closest: 220,720 miles

Spins on axis in: 27.3 days
Circles the Earth in: 27.3 days
Speed in orbit: 2,282 mph
Goes through its phases in: 29.5 days
Surface gravity (Earth's gravity =1): 1/6

People file

Name	Nationality	Dates
Aristotle	Greek	384–322 B.C.
Neil Armstrong	U.S.	born 1930
Nicolaus Copernicus	Polish	1473–1543
Eratosthenes	Greek	c. 276–194 B.C.
Yuri Gagarin	Russian	1934–1968
Charles Lyell	English	1797–1875
Isaac Newton	English	1642–1727
Ptolemy of Alexandria	Greek	c. A.D. 150
Charles Richter	U.S.	1900–1985
William Smith	English	1769–1839
James Ussher	Irish	1581–1656
Alfred Wegener	German	1880–1930

EARTH DATA

THE GREAT ESCAPE

To beat gravity and go into orbit around the Earth, you have to be launched from the ground at a speed of at least 17,360 mph. This is called the orbital velocity. But even in orbit, gravity is still trying to pull you back to Earth. To escape from gravity completely and travel, say, to the Moon, you have to be launched away from Earth at escape velocity—no less than 24,800 mph.

IN A SPIN

Because the Earth spins round in space, places on the equator travel at a speed of more than 992 mph.

THE GREATEST

The Pacific is by far the largest of the world's oceans, covering an area of more than 65 million square miles. This is greater than the area of all the land masses put together.

WAY BELOW AVERAGE

On average, the oceans are about 2½ miles deep, but some parts are much deeper. The deepest regions are the great trenches that gash many of the ocean floors. Deepest of all is the Marianas Trench in the Pacific Ocean. In 1960, the bathyscape *Trieste* dived to the bottom of the trench, registering a depth of 35,804½ feet.

THE HIGHEST PEAK

Mount Everest in the Himalayan mountain range is the highest point on Earth, with its peak 29,021½ feet above sea level. It was named after Sir George Everest, a former British Surveyor-General of India.

SHOCKING STATISTICS

Up to 100,000 earthquakes can be felt every year somewhere in the world. Of these, as many as 1,000 cause damage and death. The greatest death toll following an earthquake in modern times occurred in the Tangshan (China) earthquake of July, 1976, when about 750,000 people lost their lives.

MOON DATA

NEW ON THE MOON

An entirely new mineral in a rock sample brought back by the *Apollo 11* astronauts has been named armalcolite, after the names of the three *Apollo 11* astronauts Neil Armstrong, Edwin Aldrin, and Michael Collins.

EARTHSHINE

At the time of the crescent Moon, when only a thin sliver of light can be seen, the dark part of the Moon shows up faintly. This is called earthshine because it is caused by light from the Earth being reflected. Traditionally, this is known as "new Moon in the old Moon's arms."

BLUE MOON

"Once in a blue Moon" means very occasionally, because every now and again the Moon does turn blue. This happened in August, 1883 because of huge amounts of ash blasted into the air by the erupting Krakatoa volcano. Ash from huge forest fires in Canada produced a blue Moon in September, 1950.

BIG BAILLY

Bailly is the largest crater on the Moon, measuring about 183 miles across. It is not easy to spot because it is close to the Moon's southern limb (edge). Nearly as large and easier to see is the 143-mile wide Clavius.

MOON MADNESS

People used to believe that gazing at the Moon for too long would make you mad. This is why mad people were called lunatics, from *luna*, the Latin word for Moon.

OLD PRINTS

So little change takes place on the surface of the Moon that the footprints of the Apollo astronauts will probably still be visible in 10 million years time.

ROCK OF AGES

The oldest rock the Apollo astronauts brought back from the Moon is believed to be 4,600 million years old.

GLOSSARY

asteroids
Lumps of rock that orbit the Sun in a band between the orbits of Mars and Jupiter.

atmosphere
The layer of gases around the Earth or other planet.

axis
An imaginary line around which a body spins.

calendar
A means of dividing up the year into months, weeks, and days.

climate
The typical weather of a place during the year.

continental drift
A theory that the continents are changing their position because of movements of the Earth's crust.

core
The center part of the Earth or any heavenly body.

crater
A pit on the surface of a planet or moon produced by a meteorite or an erupting volcano.

crust
The hard, outer layer of a planet or moon.

day
The time the Earth takes to spin round once in space.

earthquake
The violent shaking of the ground when underground rocks suddenly move.

eclipse
The time when one heavenly body passes in front of another and blocks its light.

Equator
An imaginary line around the Earth, midway between the North and South Poles.

erosion
The wearing away of the landscape by natural forces, such as the weather and flowing water.

fossil
The remains of something that lived a long time ago.

geology
The scientific study of the Earth.

global warming
The gradual warming up of the Earth's climate.

gravity
The force that attracts one lump of matter to another, such as the Moon to the Earth.

igneous rock
A rock formed when molten rock cools, either on the surface or below ground.

lunar eclipse
An eclipse of the Moon.

magnetism
A force experienced by iron and a few other metals. The Earth and some other planets have magnetism.

mantle
A deep layer of rock under the crust of the Earth or other body.

mare
A sea, or flat plain on the Moon; plural *maria*.

metamorphic rock
A new rock formed when an older rock has been changed by heat or pressure.

meteorite
A lump of rock from space that falls to the surface of a planet or moon.

mineral
A chemical compound that occurs naturally in the ground.

moon
A natural satellite of a planet, such as the Earth's Moon.

moonquake
A shaking of the Moon, caused by movements of the underground rocks.

ocean ridge
A range of mountains that occurs in the middle of oceans, made by molten rock welling up from below.

orbit
The path in space a heavenly body follows when it circles around another, such as the Earth around the Sun and the Moon around the Earth.

organic compounds
Compounds containing chains of carbon atoms, widely found in living things.

ozone layer
A layer of a gas called ozone, found high in the Earth's atmosphere.

phases of the Moon
The changing shapes of the Moon in the sky during the month.

photosynthesis
The process green plants use to make their food in daylight.

planet
A large body that circles in space around the Sun, such as the Earth.

plate
A section of the Earth's crust, which moves slowly.

precipitation
Rain, snow, or other ways in which water comes out of the atmosphere and falls to the ground.

regolith
The fine soil that covers the Moon's surface.

rill
A channel on the Moon's surface in which lava once flowed.

satellite
A body or an object that circles around another body in space. The Moon is a natural satellite of the Earth. The Earth also has many artificial satellites.

sea, on the Moon
See *mare*.

sea-floor spreading
The gradual widening of an ocean due to the movements of plates of the crust.

season
A period of the year when temperatures and the general weather are much the same as in previous years.

sedimentary rock
Rock formed when ancient deposits, or sediments, are pressed together.

seismomometer
An instrument that registers shakings in the ground. A seismograph makes a record of the shakings.

solar eclipse
An eclipse of the Sun.

solar system
The Sun's family, which includes the Earth, the planets, and their moons, asteroids, and comets.

tides
The to-and-fro movements of the sea, caused mainly by the pull of the Moon's gravity.

total eclipse
The moment during an eclipse of the Sun when the Moon totally covers up the Sun and makes the sky dark.

universe
Everything that exists— space, stars, planets and all the other heavenly bodies.

water cycle
The never-ending process in which water circulates between the surface of the Earth and the atmosphere.

INDEX